D1283280

BIG JOBS, BIG TOOLS!

TOWERING CRANES

MARIE ROGERS

PowerKiDS press

New York

Published in 2022 by The Rosen Publishing Group, Inc.
29 East 21st Street, New York, NY 10010

Copyright © 2022 by The Rosen Publishing Group, Inc.

All rights reserved. No part of this book may be reproduced in any form without permission in writing from the publisher, except by a reviewer.

First Edition

Portions of this work were originally authored by Kenny Allen and published as *Giant Cranes*. All new material in this edition authored by Marie Rogers.

Editor: Greg Roza
Cover Design: Michael Flynn
Interior Layout: Rachel Rising

Photo Credits: Cover, p. 1 potowizard/Shutterstock.com; pp. 4, 6, 8, 10, 12, 14, 16, 18, 20, 22 (background) 13Imagery/Shutterstock.com; p. 5 bogdanhoda/Shutterstock.com; p. 7 Arterra/Contributor/Universal Images Group/Getty Images; p. 9 Bloomberg/Contributor/Getty Images; p.11 Image Source/Getty Images; p. 13 Pla2na/Shutterstock.com; p. 15 PHILIPPE HUGUEN/Staff/AFP/Getty Images; pp. 17, 19 Mr. Amarin Jitnathum/Shutterstock.com; pp. 21 Ben Birchall - PA Images/Contributor/PA Images/Getty Images.

Library of Congress Cataloging-in-Publication Data

Names: Rogers, Marie, 1990- author.
Title: Towering cranes / Marie Rogers.
Description: New York : PowerKids Press, [2022] | Series: Big jobs, big tools! | Includes bibliographical references and index.
Identifiers: LCCN 2020021710 | ISBN 9781725326798 (library binding) | ISBN 9781725326774 (paperback) | ISBN 9781725326781 (6 pack)
Subjects: LCSH: Cranes, derricks, etc.–Juvenile literature.
Classification: LCC TJ1363 .R593 2022 | DDC 621.8/73–dc23
LC record available at https://lccn.loc.gov/2020021710

Manufactured in the United States of America

Some of the images in this book illustrate individuals who are models. The depictions do not imply actual situations or events.

CPSIA Compliance Information: Batch #CSPK22. For Further Information contact Rosen Publishing, New York, New York at 1-800-237-9932.

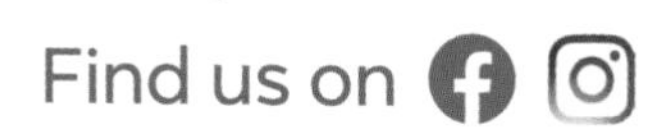

CONTENTS

Way Up There!

Cranes come in many sizes. They help lift heavy **loads**. You may have seen a crane at a **construction** site. Some cranes are on trucks. The strongest cranes help build **oil platforms**. The tallest build skyscrapers!

Tower and Jib

Cranes have tall **frames**. This part is also called the tower. The tower reaches high into the sky. Many large cranes also have a jib. A jib reaches out over the ground. Together, the tower and jib help raise heavy loads.

jib
tower

One, Two, Three...Lift

Cranes use cables, pulleys, and hooks. Pulleys are wheels that hold cables in place. Pulleys also make heavy loads easier to lift. The cable may have a hook at the end. Hooks hold on to heavy loads.

Lifting Power

A hoist helps raise and lower loads. A crane's cable is wrapped around a part called the **drum**. The hoist spins the drum to let out cable. To lift something, the hoist's drum spins the other way.

Towering Above

Workers use tower cranes to build skyscrapers. These cranes have long jibs to help lift and move loads. When a tower crane needs to reach higher, workers add a new section to the tower. The crane grows with the skyscraper!

Cranes on Trucks

Truck cranes can be moved easily. The frame of a truck crane is called the boom. The tallest truck crane can lift loads 50 stories! They are used to set up wind **turbines**.

Overhead

Overhead cranes are often used in factories. They don't have towers. The hoist moves on an overhead beam. The beam moves on a set of tracks. The crane can pick something up and move it across the building.

Giant Cranes!

Giant overhead cranes are called gantry cranes. These huge machines lift shipping **containers** off ships and put them onto trucks. The hoist moves back and forth on a beam. The entire crane moves on wheels or tracks.

"Big Carl"

The crane SGC-250 (nicknamed "Big Carl") is the biggest land crane in the world. It helps put together very large buildings. It can lift loads 820 feet (250 m) into the air. It can lift 5,512 tons (5,000 mt)!

More About the SGC-250

- Moves on 3.7 miles (6 km) of steel track
- Number of wheels: 96
- Number of engines: 12
- Fun Fact! The "Big Carl" is named after an owner of the company that built it, Carl Sarens.

GLOSSARY

construction: Having to do with the act of building something.

container: An object used to hold something else.

drum: A large spool used to wind up rope or cable.

frame: Something that holds up or gives shape to something else.

load: Something to be lifted or carried.

oil platform: An oil drilling rig at sea.

turbine: An engine with blades that are caused to spin by pressure from water, steam, or air.

FOR MORE INFORMATION

WEBSITES

How Does a Crane Work?
www.wonderopolis.org/wonder/how-does-a-crane-work
Check out this website to learn even more about towering cranes.

How Tower Cranes Work
www.science.howstuffworks.com/transport/engines-equipment/tower-crane.htm
Learn more about tower cranes and see photographs of them at work.

BOOKS

Zachary, Paul. *Cranes*. Hallandale, FL: Ez Readers, 2020.

Zalewski, Audry. *Cranes.* Minneapolis, MN: ABDO, 2019.

Publisher's note to parents and teachers: Our editors have reviewed the websites listed here to make sure they're suitable for students. However, websites may change frequently. Please note that students should always be supervised when they access the internet.

INDEX